Jhosefe Bruning
Adroaldo Dias Robaina
Marcia Xavier Peiter

Energy Assessment in Irrigation

AF294817

Jhosefe Bruning
Adroaldo Dias Robaina
Marcia Xavier Peiter

Energy Assessment in Irrigation

Conventional sprinkler application rate

ScienciaScripts

Imprint

Any brand names and product names mentioned in this book are subject to trademark, brand or patent protection and are trademarks or registered trademarks of their respective holders. The use of brand names, product names, common names, trade names, product descriptions etc. even without a particular marking in this work is in no way to be construed to mean that such names may be regarded as unrestricted in respect of trademark and brand protection legislation and could thus be used by anyone.

Cover image: www.ingimage.com

This book is a translation from the original published under ISBN 978-613-9-61585-8.

Publisher:
Sciencia Scripts
is a trademark of
Dodo Books Indian Ocean Ltd. and OmniScriptum S.R.L publishing group

120 High Road, East Finchley, London, N2 9ED, United Kingdom
Str. Armeneasca 28/1, office 1, Chisinau MD-2012, Republic of Moldova, Europe
Printed at: see last page
ISBN: 978-620-7-87059-2

Copyright © Jhosefe Bruning, Adroaldo Dias Robaina, Marcia Xavier Peiter
Copyright © 2024 Dodo Books Indian Ocean Ltd. and OmniScriptum S.R.L publishing group

Table of contents:

SUMMARY
ENERGY ASSESSMENT IN IRRIGATION

Rio Grande do Sul is experiencing a growing increase in agricultural production, however, the occurrence of phenomena that cause anomalies in rainfall negatively affects agricultural development. Faced with this situation, irrigation stands out as a possible solution for agriculture. On properties with areas of up to 10 ha, the use of conventional sprinkler irrigation is a viable alternative. However, this is responsible for a large part of the energy consumption in rural areas, due to the lack of control over the flow required to supply the area to be irrigated. This is due to the lack of control or technical basis regarding the actual volume distributed per unit of time. With this in mind, the aim of this work is to analyze the performance of different flow control methods in conventional sprinkler irrigation systems, through simulations of different system configurations. This is justified by the fact that electricity consumption is one of the main factors responsible for the increase in production costs in irrigated agriculture and also by the real possibility of a crisis in the energy sector. In view of this situation, different methods of flow control in irrigation systems can be cited, such as those that do not alter the rotational speed and those that alter the rotational speed of the motor-pump set in the pumping system. In this way, irrigation systems were designed to cover irrigated areas varying in size from 1, 3, 5, 7 and 10 hectares, comprising different combinations of spacing between sprinklers and irrigation branches. The following spacings were considered: 12 x 12 m; 12 x 18 m; 18 x 18 m; 18 x 24 m and 24 x 24 m, where different application methods were applied: no flow control and fixed irrigation time - M0 (reference method); no flow control and variable irrigation time - M1; control by throttling the valve and fixed irrigation time - M2; variable electrical frequency and fixed irrigation time - M3 to estimate energy consumption. However, it was possible to reveal the importance of using a flow control method in conventional sprinkler irrigation system projects in order to improve energy efficiency, with values ranging from 9.12% to 50.45% in the reduction of energy consumption. In addition, it can be concluded that the method that adjusts the flow rate by varying the rotation of the pumping systems was the one with the best results.

Key words: Energy saving. Flow control. Efficiency energy

Chapter 1

1 INTRODUCTION

In recent decades, Brazilian agriculture has seen a growing increase in agricultural production. This is due to new socio-cultural patterns and economic formats brought about by the need to increase food production as a result of the population's growing need for food and also the increase in population.

However, due to the large number of factors involved, agriculture is an activity that presents a high economic risk compared to other ventures. Therefore, in order to guarantee and increase crop productivity, there is a need to use techniques and technologies to reduce investment risks.

Among the risks inherent in agricultural production, those related to the climate stand out. In the state of Rio Grande do Sul, the probability of droughts damaging agriculture is seven out of every ten years (SEAPA, 2012). According to Gross and Cassol (2015), one of the factors contributing to the occurrence of droughts in the state is the climatic phenomenon known as La Niña, which causes a reduction in rainfall. This phenomenon has a negative impact on agricultural development in the affected region. In view of the aspects presented, the need for supplementary irrigation as an alternative to improve agricultural production throughout the year is evident, as well as guaranteeing food security. As a result, one solution found in the state was the creation of programs to encourage the implementation of irrigation systems as a tool to mitigate the damage caused by droughts.

However, according to Mantovani et al. (2009), irrigated agricultural production is a major user of water resources. In addition, these activities have a high water consumption among the processes that use this natural resource. Consumption varies according to the efficiency of the irrigation system used. Usually the methods with the lowest water application efficiency are related to surface irrigation and its variants. In this class of system, it is estimated that efficiency ranges from 23% to 77.8% (RODRÍGUEZ GONZÁLEZ et al., 2013).

According to Paulino et al. (2011), in the state of Rio Grande do Sul, the largest irrigated area is cultivated with rice, which occupies 82.6% of the areas with flood irrigation systems. In second place are sprinkler irrigation systems (central pivot and conventional sprinkler) whose application efficiency is greater than that of flood irrigation. According to Schons et al. (2014), these application efficiency rates for the central pivot type system are 85.89% to 87.40% for two regions in the state of Rio Grande do Sul, and these values are considered good according to the requirements of technical standard NBR 14244.

Irrigation systems that are inadequately sized, operate in an unregulated manner and have low application efficiency imply low energy efficiency. In addition, these situations cause irrigators to increase production costs due to wasted water resources and uneven crop production, as mentioned by Feitosa et al. (2014).

Due to the scarcity and possible conflicts over the use of water and electricity, there is a clear concern over the use of these resources. It is necessary to adopt management strategies that save money without damaging crop productivity by carrying out studies aimed at improving hydraulic and electrical performance, with a view to minimizing production costs.

According to Turco et al. (2009), in rural areas an irrigation system can be responsible for a large proportion of energy consumption. This is due to the lack of control over the actual volume distributed per unit of time, and the lack of technical sizing criteria has a direct influence on this consumption.

The use of flow control methods in irrigation systems can be a viable alternative, as they improve the efficiency of irrigation pumps and consequently improve the user's profitability. The use of flow control methods is justified by the fact that electricity costs can represent around 30% of production costs (ARAUJO et al., 2011).

The American agency NRCS (2014) and Lima et al. (2015) cite that when a pump is needed for different irrigation systems, the choice is made to meet the greatest demand in terms of flow and head required for the system to function. Thus, when the system operates partially, there can be an oversizing in terms of the electrical energy required by the motive force, causing an increase in its consumption as well as greater reactive power causing a low power factor which consequently leads to fines. As a result, there is significant potential for rationalizing the use of electricity under these operating conditions. Yacoubi et al. (2010) report that rising energy costs are an important factor in improving the performance of sprinkler irrigation systems.

Therefore, it is necessary to adopt techniques that can help improve the use of water and electricity in order to produce sustainably, making better use of resources and reducing energy consumption in irrigation activities (TARJUELO et al., 2015; REIS et al., 2017).

With the aim of analyzing the sizing of irrigation systems and evaluating the impacts caused by the energy factor on the cost of irrigated agricultural production, studies aimed at reducing energy consumption and applying water more efficiently are important issues.

Considering the above, the general objective of this work was to analyze the performance of different flow control methods in conventional sprinkler irrigation systems through simulations of different system configurations. The specific objectives were to quantify energy consumption in the different configurations established (in terms of area, spacing between sprinklers, number and length of branches), using different flow control methods and to identify the most energy-efficient design arrangement.

Chapter 2

2 LITERATURE REVIEW

This chapter is dedicated to a review of the main subjects that underpinned this work, through which it was possible to study the characteristics of the irrigation system and its management practices, as well as obtaining information related to the methods tested.

2.1 IRRIGATION REQUIREMENTS

In the state of Rio Grande do Sul (RS), one of the main factors in agricultural production is related to crop water requirements, which is one of the factors responsible for the frequent oscillation in crop yields.

The productive potential of agricultural crops in various regions is often limited by the occurrence of droughts which directly affect agricultural production during critical periods, due to the irregularity of rainfall which affects crop development. As agricultural production is the mainstay of the economy in the state of Rio Grande do Sul, agricultural productivity is largely responsible for determining the continuity and financial growth of farmers.

However, Gross and Cassol (2015) report an increase in periods with reduced rainfall, which is aggravated by the occurrence of the "LA NIÑA" phenomenon. These events constitute an obstacle to agricultural production, thus becoming one of the determining factors in crop yields.

In the search for agricultural practices that minimize the damage caused by the impact of droughts and possibly increase production, the use of irrigation can be encouraged. In this way, the main objectives of irrigation are to increase profits by increasing production and quality, as well as opening up new areas for agriculture in places where it would not be possible to plant crops (VIEIRA et al., 2011; FRIGO et al., 2013; MORAES et al., 2014).

2.2 WATER PUMPING SYSTEMS FOR IRRIGATION

Water pumping systems are used to convey water from points with a lower geometric elevation to another point in a given area. Providing continuous water supply and sufficient flow must be the reason for being designed to meet the required consumption demand (WU et al., 2014; SANTOS et al., 2016).

Generally speaking, an irrigation pumping system is made up of just three elements: the suction pipe, the motor-pump set and the discharge pipe with accessories (ROBAINA, 2015).

Projects with planning and sizing flaws are common practice in pumping stations and are often developed by unqualified people. This is responsible for the low efficiency of pumping stations (ROCAMORA et al., 2013).

One of the design criteria for pumping systems that should be taken into account is the diameter of the pipes, as this affects energy consumption. Special parts and fittings installed in the piping are also elements responsible for energy losses and deserve attention (KÖPP et al., 2016).

Undersized pipe diameters lead to an increase in pressure losses due to excess velocity and, consequently, an increase in the power required by the motor-pump set.

This increase in power increases the cost of the booster set and also leads to higher energy costs. However, with the use of larger diameters, these costs behave in the opposite way, making it economically viable over the useful life of the equipment (LIMA et al., 2009).

Therefore, the sizing of pipes is a very important aspect of energy efficiency, since pressure losses in the pipe have a direct impact on the design flow rate, which in turn influences the power of the pump, resulting in greater or lesser energy consumption.

Therefore, the right choice of piping is important for energy efficiency, as it has direct consequences for the design flow rate, which directly influences the pump's power, resulting in a greater or lesser loss of energy (FORTES et al., 2012).

2.3 SPRINKLER IRRIGATION SYSTEMS

According to Paulino et al. (2011), in relation to the irrigation methods used in Brazil, it can be seen that the country's irrigated agricultural production is at a different level when compared to the rest of the countries. While much of the world uses surface irrigation methods, in Brazil pressurized irrigation is the most prominent, with the sprinkler method (without a pivot) being the most widely used, as shown in Figure 1. The same author points out that in the state of Rio Grande do Sul the irrigation system that stands out is flood irrigation, with the sprinkler method (without a central pivot) being the fourth most used in the state.

Figure 1 - Distribution of the irrigated area in Brazil using different irrigation methods

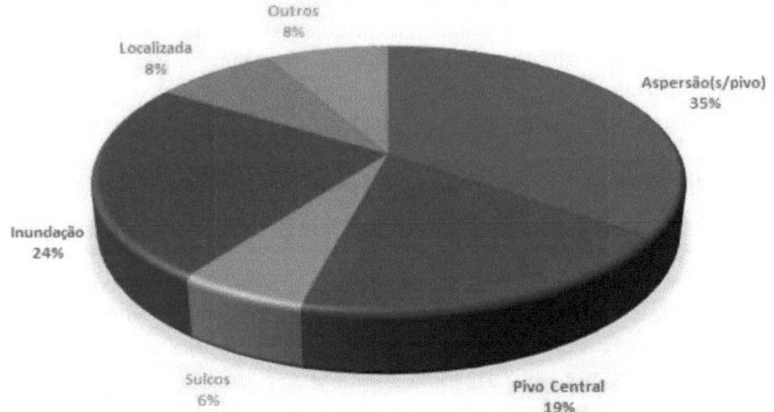

Source: Author (adapted from Paulino et al., 2011).

Sprinkler irrigation is a method in which water is sprayed over a certain area, similar to rainfall, due to the fractionation of the water jet into small drops, as the water passes through the sprinkler nozzles under a certain pressure (BERNARDO; SOARES; MANTOVANI, 2006). Sprinkler irrigation methods are divided into two irrigation systems: conventional sprinkler irrigation and mechanized sprinkler irrigation. In the former, the system is moved manually or in a fixed manner within the irrigated area. The second system uses mechanical elements to distribute the water (AZEVEDO NETO, 1998).

According to Martins et al. (2011), conventional sprinkler irrigation systems stand out for their use in small and medium-sized areas, due to their wide suitability in various locations. This system is characterized by its average energy consumption and high manpower requirements for changing lines (fixed main line and mobile lateral lines). In terms of implementation costs, it has a lower cost per unit area when compared to automated systems such as central pivot, mobile lateral and localized irrigation in all its variants (microsprinkler, traditional drip, subsurface). In addition to its lower cost compared to other systems, it is also easy to manage, which makes it the preferred system for farmers wishing to introduce irrigation on their properties and for family farming (REIS et al., 2017).

Farias et al. (2009) state that conventional sprinkler irrigation systems should be designed to apply water as uniformly as possible to the area to be irrigated, since uneven water application reduces the economic return and increases the environmental impact of irrigation, due to the drop in productivity of irrigated crops and the unnecessary consumption of energy and water.

2.4 DEMAND FOR WATER AND ENERGY RESOURCES

In the current agricultural situation, there is a growing scarcity of and conflict over the use of water and electricity, highlighting concerns about the efficient use of these resources. Martins et al. (2011) and Lima et al. (2015) report that the availability of water and energy resources has been continuously depleted, becoming an increasingly limiting factor due to inadequate use, whether due to a lack of sizing, technical criteria or even an increase in global demand.

According to Feitosa et al. (2014) electricity is one of the most important parameters for both economic growth and the social development of a region. Electricity is present in the most diverse sectors of production. In the case of water pumping systems for irrigation, in most cases there is a great dependence on electricity. This is due to the need to maintain production during periods of water scarcity and thus meet the needs imposed by the increasingly competitive consumer market.

Energy consumption in irrigation systems is increasing due to the modernization of the activities employed. Therefore, the efficient use of water and energy is taking on a major role in agriculture due to the upward trend in energy prices, making it one of the main costs for irrigators (RODRÍGUEZ DÍAZ et al., 2011; CÓRCOLES et al., 2016).

As reported by Sánchez et al. (2008) and Turco et al. (2009), irrigation is responsible for a large part of energy consumption in rural areas, and this is a considerable part of operating costs in irrigated agricultural production. This consumption can be greater when no type of control is applied during irrigation.

According to Lima et al. (2009), owners of irrigation systems do not usually seek information on the energy efficiency of their equipment. In general, owners obtain the irrigation projects developed by the suppliers of the irrigation systems. Thus, in the search for a lower initial investment, the costs of increasing the ratio of installed power per unit area are often overlooked.

The rationing of water and energy resources and their high cost in recent years has led the agricultural sector to live with a scenario where it is necessary to use artifices such as scientific forecasts, consumption analysis, planning and strategies for better use of energy. Through these studies aimed at better energy efficiency, improvements in competitiveness can be achieved by reducing costs and minimizing energy-related problems (MOUSAVI-AVVAL et al., 2011).

According to Moraes et al. (2014) and Camboim et al. (2014), excessive energy costs in irrigation systems are due to problems such as: operational   procedures; equipment

Low-performance pumping (old, outdated and poorly dimensioned technically); pipelines with excessive pressure drop; poor preventive or corrective maintenance; low efficiency of electrical or hydraulic equipment; inadequate electricity contracts and inadequate water quantity applied.

James et al. (2002) report that most water distribution systems worldwide have the potential to reduce energy consumption by at least 25% by adopting energy and hydraulic efficiency measures.

Many actions based on saving these resources are linked to pumping systems, especially with regard to irrigation systems. In this way, devices and improvements are being sought to make the use of these resources increasingly efficient in irrigation systems (FERNANDES et al., 2015). By reducing electricity costs, producers will see a return in the form of lower production costs.

Rodríguez Diaz et al. (2009), Cobo et al. (2011) and Navarro Navajas et al. (2012), refer in the literature to the possibility of sectorization in irrigation projects, highlighting the possible positive impacts and potential to obtain more effective results in terms of energy savings through the use of sectorized irrigation systems.

Studies carried out in Spain have shown significant economic gains in irrigation systems after energy audits were carried out, which resulted in greater energy efficiency of the systems as a whole (MORENO et al., 2010).

2.5 FLOW CONTROL METHODS

In the irrigated agriculture sector, electricity costs account for around 30% of production costs (ARAUJO et al., 2011). These figures are often linked to the inefficiency of the irrigation system due to sizing, equipment and even lack of system management, resulting in energy costs (MESQUITA, 2009).

In addition, the waste of electricity is at a very high level, where every year in Brazil, there is a loss of electricity equivalent to the production of two Itaipu plants, and in 2006 the loss of electricity was in the order of 38 billion kWh, which corresponds to 10% of domestic consumption (ZOCOLER et al., 2008).

Thus, the use of strategies that can reduce or control energy consumption are alternatives that make it possible to reduce energy-related costs. Therefore, it is necessary to analyze the performance of pumping stations in order to determine management strategies to increase energy efficiency (CÓRCOLES et al., 2015). In order to meet these strategies for flow control in irrigation systems, it is necessary to have knowledge of certain devices or techniques, which can be mechanical, electrical or electronic.

According to the American agency NRCS (2014), which highlights the different flow control methods that can be used to apply water to the various positions of the irrigation system, we can highlight: a) methods that do not vary the speed of the hydraulic pump or the drive motor; b) methods that vary the speed of the hydraulic pump and not the electric motor; c) methods that vary both the speed of the hydraulic pump and the electric motor.

The same body, in the form of flow control without varying the speed of the hydraulic pump and motor, can mention: a) control by throttling the valve; b) excess return to the spring (recirculation or bypass); c) use of a flow regulating valve in the main line; d) use of flow regulating valves in the secondary line.

The use of valves is the most common method of varying the flow rate in small-scale irrigation systems. In most applications, control of the flow rate is achieved through the use of a valve in the suction line, pressure line or bypass line without the use of pump impeller speed control, (ROCAMORA et al., 2013).

According to Camoirano and Dellepiane (2005), pumping systems can be adapted to energy conservation projects. Conventional irrigation systems use methods to control the rate of application, the most common method being the use of valves to control the flow of the system, without changing the rotation of the motor-pump set. In this way, the system operates outside the designed working point, thus modifying the load curve, generating a large waste of electricity and reducing the useful life of the equipment (ZOCOLER et al., 2012; ARSLAN; SAHIB., 2016).

Types of flow control that vary the speed of the hydraulic pump and do not vary the motor include: a) a combination of pulleys and belts and b) other similar methods (CARVALHO; OLIVEIRA., 2008).

The type of flow control that varies the speed of the hydraulic pump and motor includes the use of electric current frequency inverters, applied to pumping systems according to Bridi (2013).

The motors are sized to meet the maximum power required. Thus, when operating at the maximum constant rotational speed, they supply the maximum designed load. However, by reducing the load, significant energy savings can be achieved when the motor's rotational speed is reduced to match the demanded load (SAIDUR; MAHLIA., 2010).

According to Sobhy et al. (2011), Moraes et al. (2011), Lamaddalena and Khila (2012) and Lima et al. (2015), the use of frequency inverters improves the operational performance of the motor-pump set used in central pivot irrigation. This use adjusts the rotation of the system according to the topographical conditions of the site, and consequently implies a reduction in energy consumption by taking advantage of the head resulting from the difference in slope level. Köpp (2015) reports that the use of frequency inverters presents itself as a great possibility for insertion in rice farming, however, the low quality of the energy supplied in certain locations limits the operation of the equipment, since it requires voltage stability for operation.

2.5.1 Flow control without varying the speed of the hydraulic pump and motor

The throttling valve control method consists of controlling the system's flow by partially opening and closing a valve installed in series with the pump. This type of control is possible with centrifugal pumps because they have an inherently higher discharge pressure limit, so the pump can be throttled at discharge without suffering damage (AMERICO, 2004).

Guimaraes (2008) mentions that by progressively closing the throttling valve, the system's flow rate is reduced and, consequently, the pressure gradually increases, which becomes an additional pressure drop.

This method is commonly used in processes where flow control is required, and is slightly more energy-efficient than by-pass control methods.

The method known as "by-pass" detour has the simple principle of making part of the fluid from the pump discharge return to the suction. This detour process is carried out via a control valve, although it is quite widespread in chiller water pumps in commercial buildings. By-pass control is also common in positive displacement pumps (AMERICO, 2004).

Arslan and Sahib (2016) report that the most efficient method found in their study was through the use of a by-pass, which provided savings of 66%, followed by the valve throttling method. However, the same authors point out that the use of these methods is only viable when there is no possibility of using methods that alter the rotation of the system.

2.5.2 Flow control with hydraulic pump speed variation and without motor variation

Belt transmission is one of the most widely used techniques in the industrial sector for coupling motors and machines. This is because it is easy to adjust the speed of the motor to that of the machine or, in the case of irrigation, to hydraulic pumps, allowing different connection geometries between the motor and the respective load (GUIMARÃES, 2008). The change in rotational speed of the driven machine occurs through the combination of pulley sizes.

However, this type of coupling has considerable energy losses, which can be improved by using belts that are more efficient due to their shape. Efficiency is related to the type of belt used, and is also related over time to slippage on the pulleys and the wear and tear caused to the equipment by their use.

This type of method works in such a way that the motor remains at a constant speed and, through the coupling, changes the rotational speed of the pump. In this way, increasing the load on the coupling output will increase the motor's torque load (SAIDUR et al., 2012).

However, this type of control does not reduce energy consumption, it is simply a way of adapting to meet a specific need (CAMPANA et al., 2000).

2.5. 3 Flow control with variable speed hydraulic pump and motor

According to the NRCS agency (2010), when it comes to energy consumption in pumping systems, the best method of reducing energy consumption is through variable speed control of the pumps in order to meet variable demands, since a slight reduction in speed can result in a significant reduction in the system's power input.

According to Panesi (2006), frequency inverters are electronic devices that act on the frequency of the electric current in motors, varying their speed. Considering that hydraulic pumps operate with varying flow rates and that these are linearly related to the speed of the machine, the use of inverters makes it possible to adjust the flow rate by changing the speed of the system.

According to Bezerra and Silva (2009), the frequency inverter consumes between 4 and 10% of the energy absorbed to drive the motor pumps. These authors mention that the inverter's performance decreases as the output frequency decreases and as the driven load decreases. When applied correctly, these losses become insignificant compared to the reduction in energy consumption from operating the pump with variable speed. In the majority of water lifting systems that require flow control, the same authors state that the most efficient speed control is by controlling the pump's rotation speed.

According to Aquino et al. (2008), in water pumping where variable speed is required, the use of frequency inverters has great potential for saving both water and energy. They also report that the benefits obtained from using inverters go beyond reducing energy consumption, such as increasing the useful life of the system's equipment, as the application of this equipment allows precise control of the flow rate, providing the same effect as mechanical strategies.

As mentioned by Lima et al. (2015), the use of a frequency inverter has proven to be a viable, effective device capable of varying the rotation according to the pressure requirements established at points considered critical in the irrigated area. It also improves the operational performance of the motor-pump set.

Therefore, in order to lend reliability and credibility to the research topic proposed in this project, following the bibliographical review, some relevant works on the subject will be condensed.

Moraes et al. (2014), studying the reduction of electricity costs for irrigation using frequency inverters in a prototype center pivot, found results for different lateral line slopes (0; 10; 20 and 30%), which were able to provide electricity savings of approximately 48; 37; 26 and 16 % for the respective slopes.

Work carried out by Araújo et al. (2004) evaluating the application of a frequency inverter in a sprinkler irrigation system with 4 lines managed alternately, where different openings were made for 1, 2, 3 and 4 lines simultaneously. The authors found that the system with a frequency inverter reduced energy consumption by around 58% and that the sprinkler's working pressure was maintained, with only the required flow rate varying according to the management of each of the lines.

Oliveira Filho et al. (2000) evaluated the behavior of the load index, performance, power factor and rotation of electric motors used in central pivot irrigation systems, with and without the use of frequency inverters. They concluded that with the use of the inverter, performance increased by around 6%. On the other hand, the power factor can be corrected to values close to unity, thus not requiring the use of a capacitor bank to make the correction. Thus, the use of this equipment is an alternative for rationalizing the use of electricity in areas with different slopes and central pivot irrigation systems.

Fortes et al. (2012) show the behavior of the characteristic curve of a pump with the application of two flow control methods used in water pumping systems. The use of throttling valves and the use of a frequency inverter as shown in Figure 2, where region A of the graph represents lower energy consumption through the use of the frequency inverter compared to the use of throttling valves.

Figure 2 - Comparison of pump curve behavior using flow control methods (throttle valve and frequency inverter)

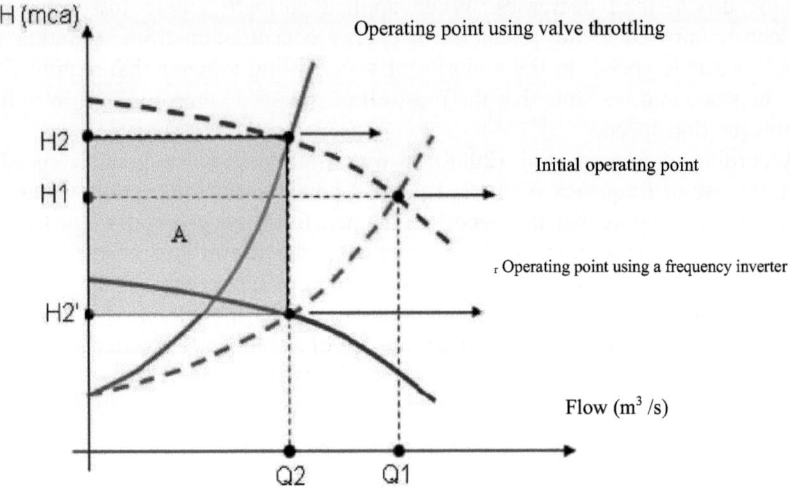

Source: Fortes et al. (2012).

In light of the above, it can be seen that the use of energy reduction methods is of great importance for use in irrigation systems.

Chapter 3

3 MATERIAL AND METHODS

This chapter presents the materials and methods used to carry out all the stages of the proposed study. In this way, the procedures used in the experiment, which were developed theoretically, will be explained.

3.1 DEVELOPMENT OF IRRIGATION PROJECTS

The irrigation projects used in this study were developed in a theoretical way, meeting the basic requirements of an irrigation project. In addition, the option of doing it in a theoretical way was to take into account the different sizes of areas along with different sprinkler spacings, lengths and numbers of irrigation branches. For all the situations developed, the flow control methods were applied. The Excel spreadsheet and the Autocad program from Autodesk® were used to carry out the sizing.

3.2 DATA COLLECTION

In this study, the conventional mobile sprinkler system was used to design different irrigation system projects for small areas.

To design the irrigation systems, reference evapotranspiration (ETo) data was collected for the municipality of Santiago.

This study considered the irrigation of annual summer pasture, using crop coefficients and the methodology proposed by Doorenbos and Kassam (1994) for water management throughout the cycle.

The irrigation rate to be applied was determined in such a way as to meet at least the criterion of being equal to or less than the real water capacity in the soil (RWC). For the purposes of calculating the net water quantity, the application efficiency set for the irrigation systems was 85%, which is considered adequate for sprinkler irrigation systems (BERNARDO; SOARES; MANTOVANI, 2006).

To determine the possible irrigation shift between irrigations, data on the actual soil water capacity (ARC) and crop evapotranspiration (ETc) were used. Based on this data, the application rate needed to meet demand was determined. A maximum watering shift of 7 days was obtained, and this value was adopted based on the data calculated above. It was considered that a worker should have one day off a week, and for the purposes of the project, 6 working days were set aside for irrigation.

The daily irrigation time was set at no more than ten hours, due to legal labor issues. Similarly, the minimum working time was set at approximately eight hours a day, because the longer the water application time for the same area, the lower the flow required by the system and, consequently, the lower the power required by the motor pump.

In order to make better use of resources and also reduce the cost of the irrigation system's piping and motor-pump set, the systems were divided into irrigation sectors, with different numbers of lateral lines operating simultaneously. In this way, the entire intended area can be covered according to the aforementioned criteria.

3.3 IRRIGATION SYSTEM CONFIGURATIONS

The conventional sprinkler irrigation systems were sized to cover irrigated areas varying in size from 1, 3, 5, 7 and 10 hectares. Thus, the system's irrigation line distributions were configured to use different sizes of areas, comprising different combinations of sprinkler spacing and irrigation branches. The sprinkler and branch spacings considered were 12 x 12 m; 12 x 18 m; 18 x 18 m; 18 x 24 m and 24 x 24 m, distributed in such a way as to cover the areas mentioned above.

The sizing of the different irrigation systems considered the system's main line positioned centrally in the area to be irrigated. Figure 3 shows a schematic representation of the standard irrigation system containing the lateral lines on both sides of the main line, as well as its basic elements.

Figure 3 - Schematic representation of an irrigation system

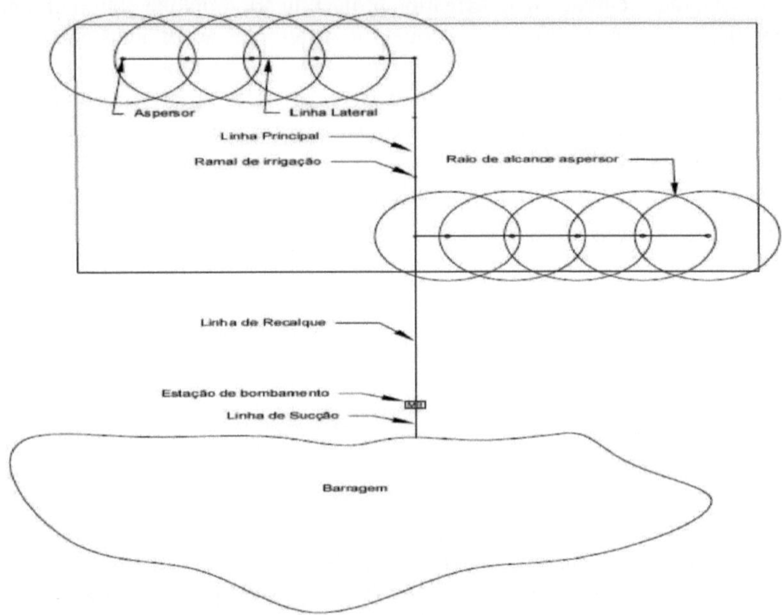

Source: Author's collection.

The sprinklers used have different nozzle diameters to meet the water requirements according to the spacing between the sprinkler and the irrigation branch. Table 1 shows the technical characteristics of the models used.

Table 1 - Technical characteristics of the sprinklers selected and used in the irrigation systems

Spacing	ASPERSOR						
	Brand	Model	s Nozzles (mm)	Flow (m^3 h)$^{-1}$	s wet (m)	Working pressure (m c a)	Application rate (mm h)$^{-1}$

12X12	Agropolo	NY 25	3,5 x 2,5	0,972	26	25	6,7
12X18	Fabrimar	Midi	4,0 x 2,8	1,41	27	25	6,5
18X18	Fabrimar	Midi	5,2 x 3,2	2,14	29	25	6,6
18X24	Fabrimar ECO	A232	5,6 x 3,2	3,08	33	40	7,1
24X24	Fabrimar ECO	A232	6,4 x 3,2	3,9	38	40	6,8

The application rate considered as the basis for the study for the different spacings was 6.48 mm h^{-1} . The sprinklers used were full-circle (360°), selected according to the spacing between sprinklers and branches. In this way, each sprinkler will deliver a certain flow rate, as shown in Table 1.

3.4 GEOMETRIC HEIGHT OF THE IRRIGATION SYSTEM

To determine the geometric heights at the insertion points of the irrigation branches in the different irrigation systems, slope values were adopted for the suction line, discharge line and main line of 4%, 7% and 6%, respectively. The irrigation lines in all the projects were considered to be level.

3.5 HEAD OF THE IRRIGATION SYSTEM

The manometric heights of the insertion points of the irrigation branches were determined by adding the pressure drop in the main line in each configuration to the respective geometric height of the insertion point.

In this study, the Hazen-Williams formula equation (1) was used to determine the main or distributed pressure drop along the main line and the irrigation lines, adopting the roughness coefficient of the PVC pipe material, where the adopted value of the material's C coefficient is 150.

The secondary or localized pressure losses in the irrigation line were determined using the method of the friction coefficients of the peps, equation (2). The pipe diameters were determined in such a way that the maximum flow velocity did not exceed 2 m s-1. For the suction pipes, commercial diameters above those defined for the respective discharge pipes were used.

$$hf = 10{,}645 \frac{L}{D^{4{,}87}} \frac{Q^{1{,}85}}{C^{1{,}85}} \tag{1}$$

$$hfloc = \sum K \frac{v^2}{2g} \tag{2}$$

Being:

Hf- Main load loss (m);

hfloc- Localized head loss (m);

L- Length (m);

D- Pipe diameter (m);

Q- Flow (m s)$;^{3\text{-}1}$

C- Hazen Willians coefficient;

K- Pepa coefficient;

v- Flow velocity (m s)$;^{-1}$

g- Acceleration due to gravity (m s)$.^{-1}$

The localized pressure drop coefficient" has values that are tabulated for the various possible types of pipe that can be used in the hydraulic piping of irrigation systems. Table 2 shows the k values for some of these pepas

Table 2 - "k" values - localized pressure drop coefficient for different pepas

Handle	K
Gradual expansion	0,3
Nozzles	2,75
Open door	1
90° elbow	0,9
45° elbow	0,4

Sieve	0,75
90° bend	0,4
45° bend	0,2
Normal entry	0,5
Edge entry	1
Check valve	2,5
Junpao	0,4
Venturi meter	2,5
Gradual reduction	0,15
Angle register	5
Drawer register	0,2
Ball valve	0,2
Output	1
Te, direct pass	0,6

Te, side exit	1,3
Te, bilateral output	1,8
Foot valve	1,75

By defining the manometric heights and flow rates required for each of the pumping systems, the power required for each of the situations established was determined, and the power required was adjusted according to the commercial power ratings as shown in Table 3.

Table 3- Definitions of the powers required and those used commercially in each of the situations studied

| Irrigated area (ha) | Power required (hp) | | | | | Commercial power used (cv) | | | | |
| | Foams | | | | | Foams | | | | |
	12x12	12x18	18x18	18x24	24x24	12x12	12x18	18x18	18x24	24x24
1	3,89	5,03	5,33	8,72	5,61	5	7,5	7,5	10	7,5
3	11,74	13,93	13,79	14,98	14,01	15	15	15	15	15
5	22,3	24,3	24,36	25,12	22,65	25	25	25	30	25
7	29,08	28,86	28,71	35,3	33,46	40	40	40	40	40

10	40,21	44,5	47,55	57,26	55,08	50	50	50	60	60

3.6 FLOW CONTROL METHODS FOR IRRIGATION WATER PUMPING SYSTEMS

With the aim of quantifying and analyzing possible reductions in energy consumption in irrigation systems, methods were applied to control the flow of water applied to the different positions of the irrigation lines. The methods used to estimate energy consumption were as follows:

a) no control and fixed irrigation time - M0 (reference method);
b) no control and variable irrigation time - M1;
c) control by throttling the valve and fixed irrigation time - M2;
d) use of frequency inverter and fixed irrigation time - M3.

The application of the flow control methods was implemented in such a way as to take into account the alteration of the characteristics that guide each of the methods. Thus, in the first two methods, the only characteristic that was altered was the irrigation system's operating time, with a single operating time being considered for method M0, while method M1 used the operating time according to the flow required for the sector to be irrigated. The third method (M2) took into account the total head required for each system and the nominal efficiency of the motor pump set to meet the necessary demand. For the M3 method, the flow rate and head for each sector were considered, taking into account the efficiency of the set.

3.7 ANALYSIS OF RESULTS

The flow control methods mentioned above altered the electrical energy consumption behavior of the irrigation systems developed. For the purposes of analyzing the results, treatment M0 was used as a control. The unit energy consumption was quantified according to the areas irrigated and the spacing between the sprinklers and branches. The comparison between the established treatments was made through energy consumption, where the savings in energy consumption between the treatments were verified.

Chapter 4

4 RESULTS AND DISCUSSION

The results are presented in this chapter in order to analyze the benefits and consequences of energy efficiency measures in sprinkler irrigation systems, where a comparison was made between the three flow control methods in relation to the uncontrolled system, considering five different area sizes and five different spacings of irrigation branches and sprinklers, in order to identify and quantify electricity consumption.

4.1 ENERGY CONSUMPTION PER UNIT AREA

In an analysis of energy consumption in irrigation systems, taking the method without flow control (M0) as a reference, it is possible to see that the behavior of energy consumption, where it was considered on a unit basis, is consumption per millimeter per unit area (kWh mm^{-1} ha^{-1}). Consumption shows a downward trend in relation to the increase in irrigated area, as shown in Figure 4, which represents the consumption ratio for each of the sprinkler spacings and branches used, in relation to the five area sizes analyzed. These results agree with the study by Viera et al. (2011), who analyzed areas from 3 to 50 ha and found that the cost of electricity showed a downward trend in relation to the increase in irrigated area.

To carry out the evaluations, the use of the same type of motor pump with identical installed powers was considered according to the needs of each area, in order to meet the correct sizing of each irrigation system. Because of this, in some situations it was not possible to use the same power as the other systems, as in the case of the 12 x 12 m spacing for 1 ha, where the power dimensioned was lower. On the other hand, for the 24 x 24 m spacing for 10 ha, the power was greater than that used in the other spacings in the same area. In addition to this, the 18 x 18 m spacing in the 1, 3 and 10 ha areas required the use of higher installed powers than the other spacings, as reported in Table 2 above. This was done in order to ensure correct sizing, so that the system was neither oversized nor overloaded.

As a result of this, it can be seen in Figure 4 that the lowest consumption occurred in the 12 x 12 m spacing to serve the 1 ha area, thus showing an opposite behavior in relation to the increase in the irrigated area. Similar behavior, but in reverse, can be seen in the 24 x 24 m spacing, where consumption tended to increase in the 10 ha area.

Energy consumption generally showed a decrease as the irrigated area increased, with the exception of the 18 x 24 m spacing, which from 3 ha onwards showed a slight increase in relation to the irrigated areas.

The data analyzed was fitted to a 2nd order polynomial equation through a regression analysis between energy consumption and the increase in irrigated area. The coefficients of determination showed that the regression equations adhered well to the data analyzed. The spacing with the best R^2 was 24 x 24 m with a value of 0.997, followed by 12 x 18 m, 18 x 18 m and 18 x 24 m with values of 0.944, 0.932 and 0.735 respectively. The 12 x 12 m spacing showed the worst performance in terms of the coefficient of determination with a value of 0.709.

Figure 4 - Behavior of unit electricity consumption (kWh mm^{-1} ha^{-1}) as a function of irrigated area (ha) for each spacing combination

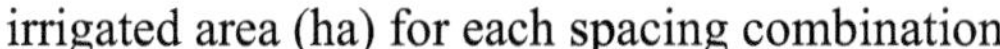

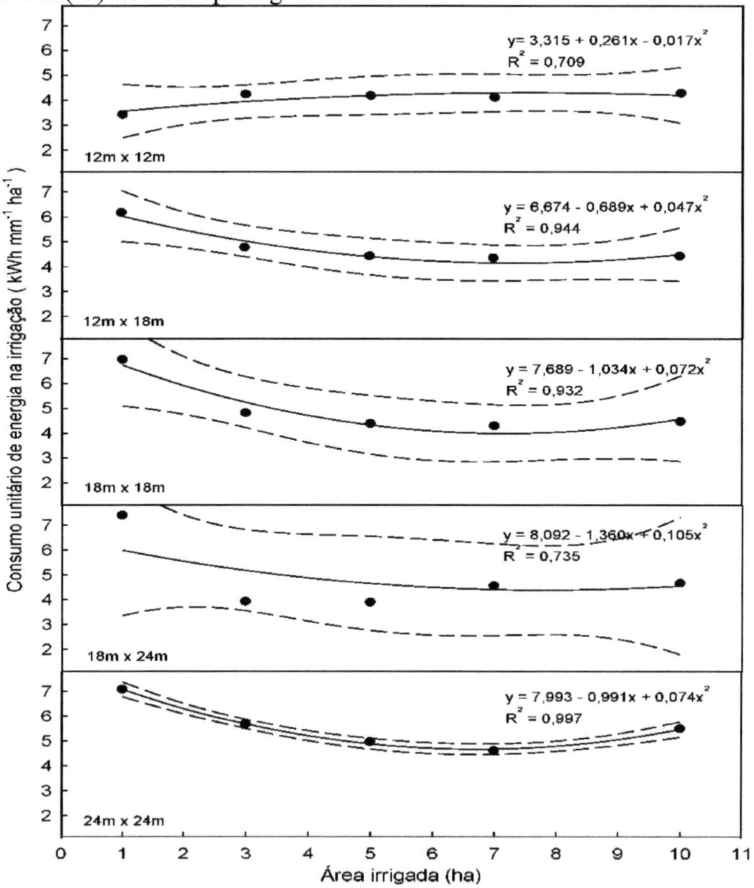

The total energy consumption required to carry out one irrigation is shown in Figure 5. It can be seen that for all the spacings adopted, the total energy consumption showed an increasing linear behaviour as the irrigated area increased, and these results are in line with those found by Rodríguez Díaz et al. (2011) in a study carried out in different regions of Spain.

The data analyzed showed a linear behavior in which the coefficient of determination values were higher than 0.97 for all the spacings studied.

Figure 5- Energy consumption required to irrigate each size of area with the M0 method

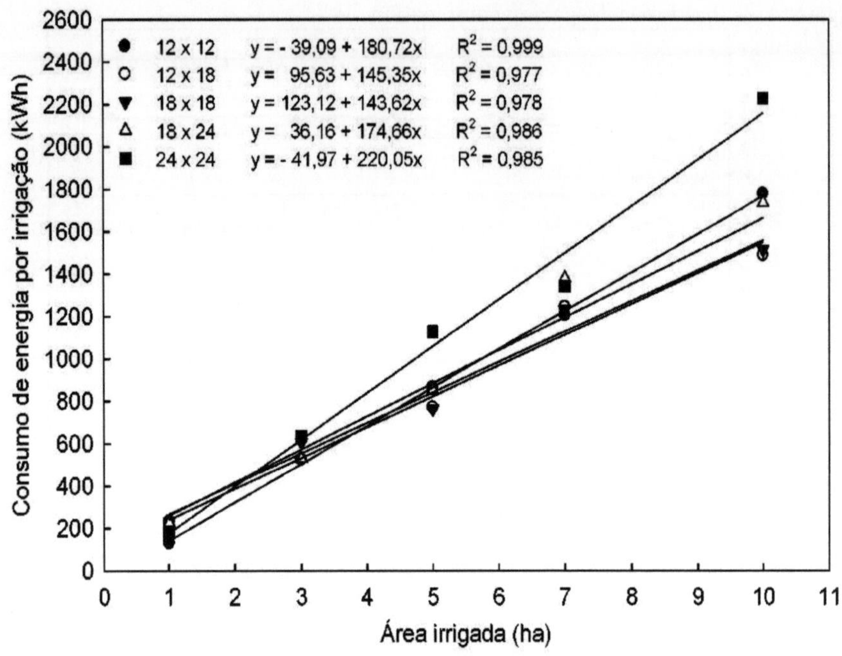

4.2 EVALUATION OF UNIT ENERGY CONSUMPTION AS A FUNCTION OF DIFFERENT SPACINGS

Electricity consumption was analyzed as a function of the different spacings between sprinklers and irrigation branches adopted. These behaviors are represented in Figure 6, which shows the variation in electricity consumption for all the area sizes studied in relation to the spacing.

It can be seen that the unit energy consumption increased when comparing the spacings, where the lowest values were found in the 12 x 12 m spacings, with values ranging from 3.4 to 4.27 kWh mm^{-1} ha^{-1} . For the 24 x 24 m spacing, the values found were 5.48 to 7.07 kWh mm^{-1} ha^{-1} .

It can also be seen in Figure 6 that energy consumption behavior was similar from 3 ha onwards, where consumption values ranged from 3.88 kWh mm^{-1} ha^{-1} for the 5 ha area to 5.67 kWh mm ha^{-1-1} for the 3 ha area. The highest energy consumption values were found for the 1 ha area.

The values found increased as the spacing studied increased, with the highest energy consumption values generally being found in the 24 x 24 m spacing.

Figure 6- Behavior of electricity consumption in relation to the spacing between sprinklers and irrigation branches in the areas studied using the M0 method.

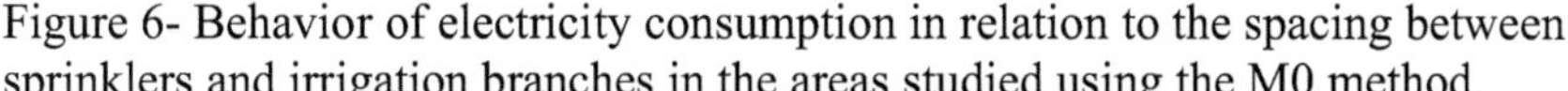

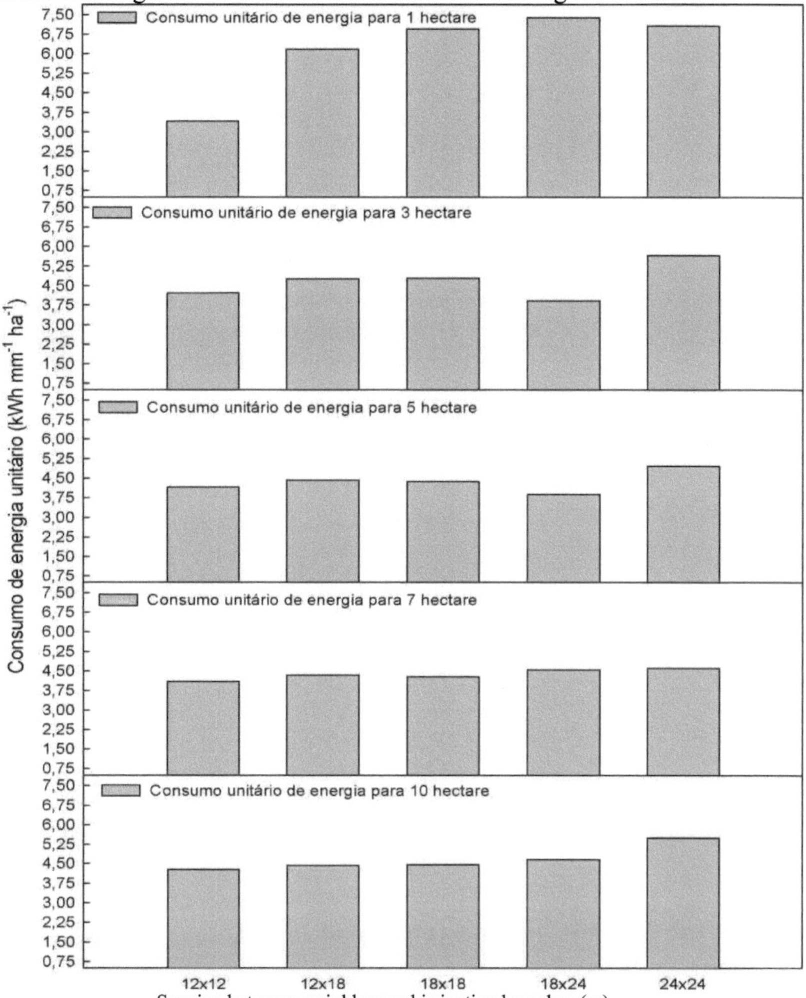

Looking at the graphs in Figure 6, it can be seen that the highest unit energy consumption occurred in the 1 ha spacing, with the exception of the 12 x 12 m spacing due to what was mentioned in relation to the motor pump set, as shown in Table 3.

Moreno et al. (2016), in a study of corn cultivation with two sprinkler-irrigated plots, found that energy consumption was slightly higher at 15 x 15 m spacing than at 18 x 18 m spacing. This result is similar to that found in this study in areas larger than 1 ha. The energy consumption values for the 18 x 18 m spacing in the 5 and 7 ha areas were lower than for the 12 x 18 m spacing. In the 3 and 10 ha areas, this behavior did not occur, however, the difference between the consumption values was minimal.

In general, it can be seen that as the spacing between sprinklers and irrigation branches increases, unit energy consumption tends to increase. This can be seen in Figure 7, which shows the average unit consumption (kWh mm^{-1} ha^{-1}) as a function of spacing.

The values found in this study are in line with those reported by Marouelli and Silva (1998), who presented energy consumption values for conventional irrigation systems ranging from 3 to 6 kWh mm^{-1} ha^{-1} . In the study carried out by Lima et al. (2009) using three central pivots, they found total energy results (kWh mm^{-1} ha^{-1}) ranging from 5.37; 8.30 to 12.40, these values being relatively higher than those found in this study. However, the standardized values for the pivots were 31.24; 12.89 and 14.91 kWh mm^{-1} ha^{-1} to provide a millimeter of water in at least 80% of the irrigated area when the geometric height is equal to 100 m.

It can be seen that the consumption of the 18 x 24 m spacing showed a small reduction compared to the 18 x 18 m spacing, thus showing a possible use of this spacing, due to a possible reduction in the number of pipes that would be used, as well as a possible reduction in costs with the use of this spacing.

Figure 7- Average unit energy consumption as a function of the spacing between sprinklers and irrigation branches for the M0 method

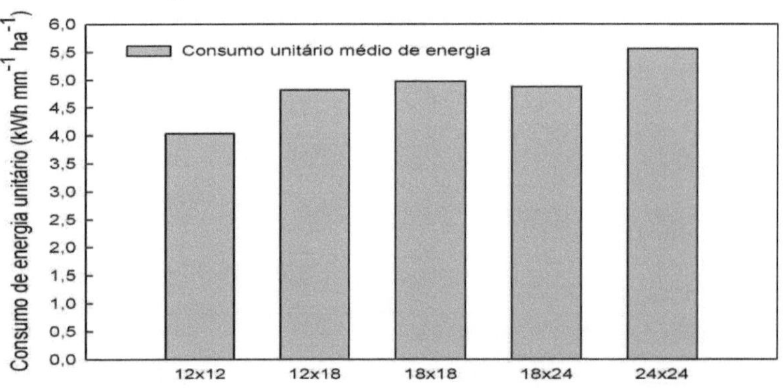

4.3 USING FLOW CONTROL METHODS TO EVALUATE ENERGY CONSUMPTION

In order to evaluate the control methods used in this study, an analysis was made in relation to the comparison with the reference method (M0). Figure 8 shows the behavior of unit energy consumption as a function of the different spacings and sizes of irrigated area.

Analyzing the 12 x 12 m spacing, it can be seen that there was a reduction in unit energy consumption, ranging from 0.78 to 1.28 kWh mm^{-1} ha^{-1} . With this spacing, energy savings tended to increase as the area increased. In the 24 x 24 m spacing, the energy saving values were higher when used in smaller areas, which is the opposite of the spacing mentioned above.

For the other spacings, the unit energy consumption values decreased as the irrigated area increased, with this behavior occurring in areas of up to 5 ha. From the 7 ha area onwards, there was an increase in energy savings, which again decreased at the 10 ha area.

Figure 8- Graph of the behavior of the energy savings obtained using the control method with varying irrigation time (M1) compared to the reference method (M0)

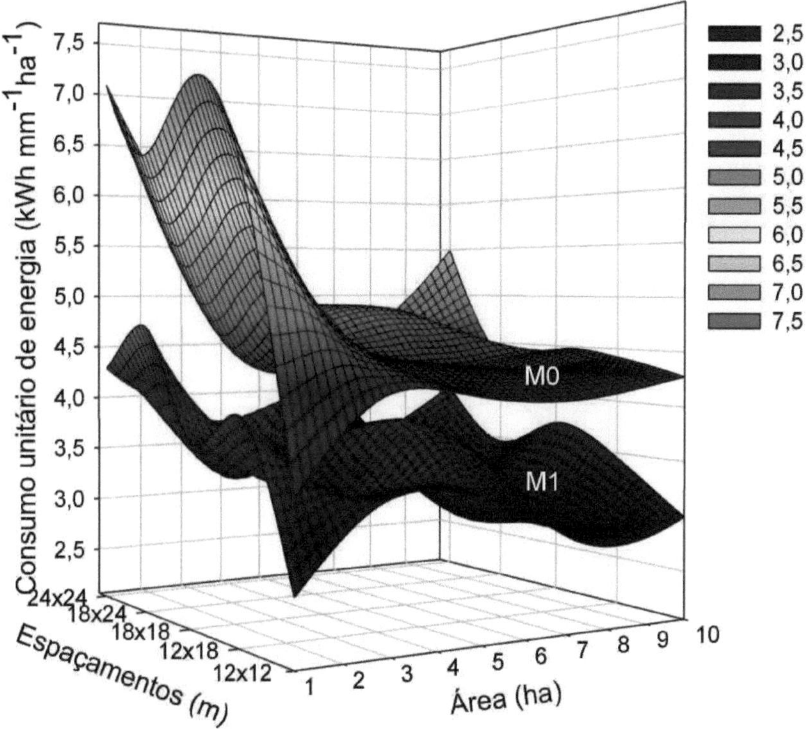

Figure 9 shows the behavior of energy consumption using the M2 method together with the M0 method as a function of area and spacing. The energy consumption values using this method showed a similar behavior to that found with the method mentioned above for the 12 x 12 m spacing, where there was an increase in savings as the area to be irrigated increased. For the other spacings, the reduction in consumption showed the opposite behavior, i.e. it decreased as the irrigated area increased.

The savings in unit energy consumption using M2 for the 12 x 12 m, 12 x 18 m, 18 x 18 m, 18 x 24 m and 24 x 24 m spacings varied from 0.70 to 1.02; 0.40 to 2.08; 0.47 to 3.18; 0.45 to 2.13 and 0.87 to 2.73 kWh mm^{-1} ha^{-1} respectively for each spacing, with the highest values found in the 1 ha areas, with the exception of the 12 x 12 m spacing.

The use of the throttling valve method as flow control shows lower energy reduction values than the other methods. This can be seen in the graphical analysis, which shows that the points are less distant from the reference control method.

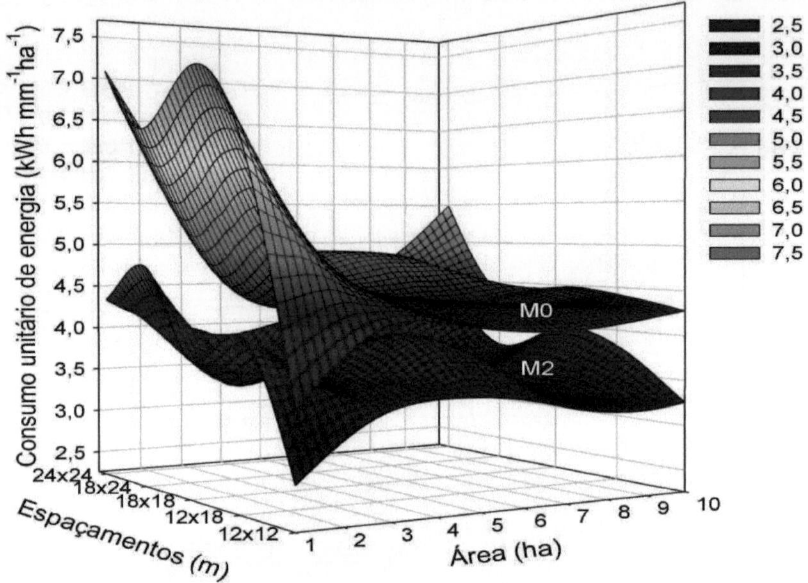

Figure 10 shows the behavior of the M3 method together with the reference method. The results found using this flow control method were similar to those reported by the methods mentioned above. However, in terms of consumption reduction values, these were higher than the other methods, with savings ranging from 0.83 to 3.50 kWh mm^{-1} ha^{-1} between the spacings, with an average value of 1.46 kWh mm^{-1} ha^{-1} in the situations analyzed in this study. As a result, it is the method that provides the greatest electricity savings when used for conventional sprinkler irrigation systems in all the situations analyzed.

Figure 10- Graph of the behavior of the energy savings obtained using the control method using a frequency inverter with fixed irrigation time (M3) in relation to the reference method.

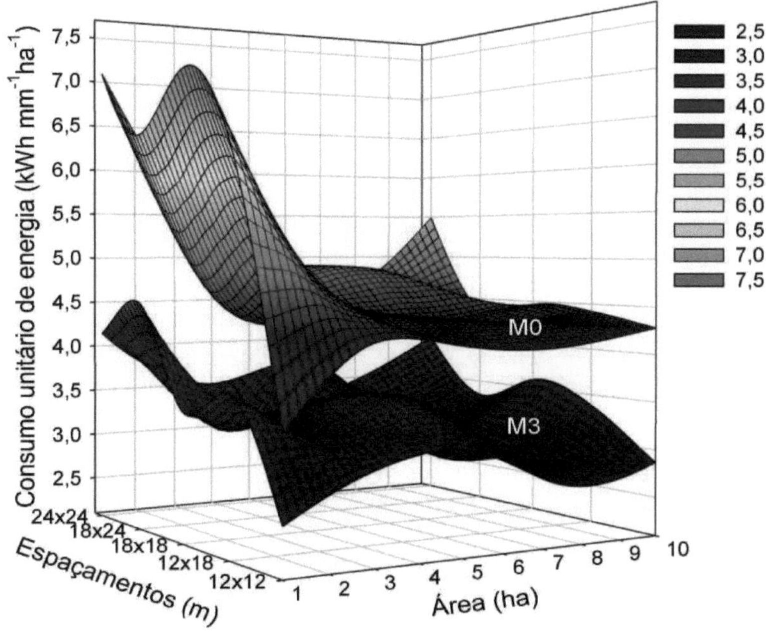

Table 4 shows the values of the coefficients of determination, standard deviation and regression adjustment parameters for each of the control methods, where it can be seen that the adjustment was to the parabolic model.

The value of the coefficient of determination for the data from the M0 method was the best, being over 60%, while for the M1, M2 and M3 methods the values were 44.6%, 43.9% and 49.8%, showing that among the control methods tested, M3 was the one with the highest value, second only to M0. It can also be seen that M3 had the lowest standard deviation among the methods tested, thus showing greater homogeneity between the values tested. The M0 method had the highest standard deviation.

Table 4 - Correlation coefficient values, coefficient of determination and equation parameters for each flow control method

Methods	R^2	dp	$Z=a+bx+cy+dx2+ey2$				
			a	b	c	d	e
M0	0,643	0,911	5,216	-0,599	0,545	0,042	-0,038
M1	0,446	0,447	3,201	-0,213	0,495	0,015	-0,055
M2	0,423	0,439	3,234	-0,158	0,440	0,013	-0,045
M3	0,498	0,349	3,057	-0,203	0,412	0,015	-0,039

Where: Z is the energy consumption variable; x is the value for the size of the irrigated area; y is the value for the area occupied by the spacing between the sprinkler and the irrigation branch and dp is the standard deviation.

Using the data found, it was possible to show how much energy consumption was reduced by adopting each of the methods. Table 5 shows the percentage reduction in energy with the use of each flow control method for each of the spacings and area sizes tested.

The lowest energy savings obtained by using the control methods mentioned above were found for areas of 5 ha for most of the spacings, with the exception of 24 x 24 m. However, the highest percentages of savings did not show a trend to be followed, as most of them were obtained in the smallest area studied (1 ha), with values of more than 20% in electricity savings.

The use of flow control methods in conventional sprinkler irrigation systems generally showed savings ranging from 9.12% to 50.45%. The method using a throttling valve showed the lowest reduction and the method using a frequency inverter showed the highest reduction in energy consumption. These values are consistent with those found by Aquino et al. (2008), who tested different percentages of valve throttling and compared them with the use of frequency inverters, and obtained similar energy reduction values ranging from 6%, 39% and 74%, which confirms the findings of this study.

Table 5 - Values for energy savings (%) using the three flow control methods adopted according to the different areas and spacings analyzed

Area (ha)	Spacing —	Flow control methods		
		M1	M2	M3
	12x12	22,80	20,52	24,30
	12x18	34,81	33,74	39,81
1	18x18	46,33	45,72	50,45
	18x24	30,30	28,81	32,14
	24x24	39,19	38,57	41,29
	12x12	20,43	20,17	28,50
	12x18	31,28	24,40	29,56
3	18x18	24,53	24,15	29,51
	18x24	18,93	13,66	19,88
	24x24	32,33	29,79	36,32
	12x12	19,84	17,22	24,09
	12x18	12,15	9,12	18,71
5	18x18	13,66	10,73	20,32
	18x24	19,04	11,59	18,58

		24x24	32,97	27,22	33,72
		12x12	28,77	19,80	31,75
		12x18	29,32	21,04	31,62
	7	18x18	30,99	22,99	33,58
		18x24	28,94	21,39	28,94
		24x24	27,18	18,92	26,72
		12x12	29,92	23,77	32,91
		12x18	21,86	13,57	23,44
	10	18x18	16,49	11,25	19,47
		18x24	20,18	13,16	21,15
		24x24	27,26	18,95	27,43

Being

M1 - no flow control with variable irrigation time;

M2 - control by throttling the valve with fixed irrigation time;

M3 - control method using a frequency inverter with fixed irrigation time.

Regarding the use of frequency inverters in irrigation systems, Moraes et al. (2014), in a study with a prototype central pivot, simulating real conditions, found electricity savings of approximately 48%, 37%, 26% and 16% for the conditions with the lateral line inclined at 0%, 10%, 20% and 30%. This value is consistent with the results found in this study, however for the level condition the value is higher than that found in this study.

Studies carried out in Italy on pumping stations have shown savings in energy consumption of 27% and 35% with the use of variable rotation (LAMADDALENA; KHILA., 2012), which are closer to the values found in this study. Similar results were found in a case study carried out at the CIHEAM-IAMB institute in Italy. Under different conditions, using inverters to change the rotational speed resulted in energy savings of around 33% (BARUTQU et al., 2007), which is in line with the results found in this study.

Results similar to those found in this work are in line with Dutra and Altafini (2014), who obtained an estimated reduction of
37.7% through a proposal to replace pumping sets and use a frequency inverter. This also agrees with a study by Haddad et al. (2010) which found that the use of frequency inverters reduced energy consumption by up to 32% when compared to operating at the nominal speed of the pumping system.

In addition, the use of an inverter helps to adjust the frequency of electric motors in order to keep the efficiency as close to nominal as possible, i.e. the efficiency of the motor operating under the condition equivalent to 100% of the nominal load. It can also help to increase efficiency (OLIVEIRA FILHO et al., 2000). The benefit of installing variable speed drives in pumping systems, in addition to saving energy, also favors greater system reliability and better process control (TUTTEROW; MCKANE, 2004).

Figure 11 shows the energy saving amplitude values for method M1 for each of the spacings tested. The use of this control method showed results for the range of energy consumption savings, for the 12 x 12 m spacing, which varied from 19.84% to 29.92%, the highest percentage achieved in the largest study area, for the 12 x 18 m spacing the range was between 12.15% and 34.81%, for the 18 x 18 m the range was between 13.66% and 46.33%, for the 18 x 24 m the range was between 18.93% and 30.30% and for the 24 x 24 m spacing the range was between 27.18% and 39.19%.

Based on these results, the spacing that showed the highest savings amplitude values can be seen.
It can be seen that the greatest range of savings occurred with the use of the 18 x 18 m spacing, where the range of values was over 30%. In addition, as previously reported, the greatest savings found using this control method were in the 24 x 24 m and 18 x 18 m spacings respectively.

Figure 11 - Amplitude of energy savings obtained using the control method with varying irrigation time (M1) as a function of the spacings analyzed.

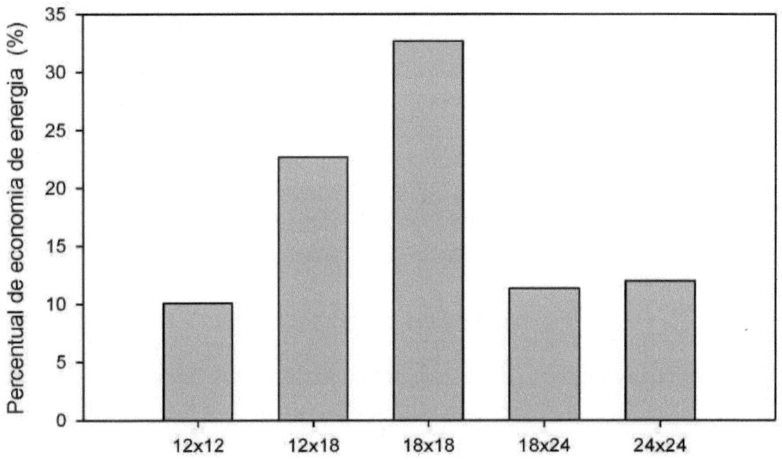

Figure 12 shows the range of savings using the M2 flow control method. The results found for the energy saving amplitudes obtained for the 12 x 12 m, 12 x 18 m, 18 x 18 m, 18 x 24 m and 24 x 24 m spacings range from 17.22% to 23.77%, 9.12% to 33.74%, 10.73% to 45.72%, 11.59% to 28.81% and 18.92% to 38.57% respectively.

Through this analysis, the 18 x 18 m spacing was again the one that showed the greatest range of reduction in energy consumption compared to the other spacings analyzed, with a value of 34.99% reduction in energy consumption. Although this method showed the greatest savings, it was the one that showed the lowest savings in all the situations analyzed in this study compared to the other methods.

Figure 12- Amplitude of energy savings obtained by using the throttling valve control method with fixed irrigation time (M2) as a function of the spacings analyzed.

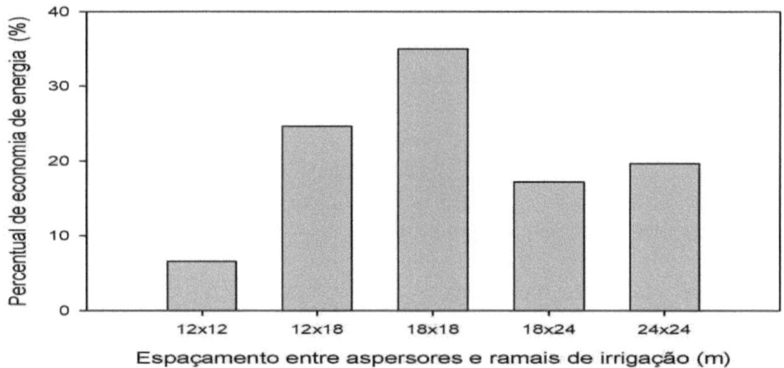

Figure 13 shows the amplitude values between the highest and lowest reduction in electricity consumption obtained by using the IF-TIV method in each space. The use of this control method showed savings amplitudes ranging from 18.71% to 50.45%, making it the method with the highest savings values in all the situations analyzed when compared to the other methods analyzed.

The results found for the range of savings for each spacing varied from 24.09% to 32.91%, 18.71% to 39.81%, 19.47% to 50.45%, 18.58% to 32.14% and 26.72% to 41.29% for the 12 x 12 m, 12 x 18 m, 18 x 18 m, 18 x 24 m and 24 x 24 m spacings respectively. Once again, the 18 x 18 m spacing was the one with the greatest range of energy consumption reduction, with a reduction of over 30%. With the use of a frequency inverter, the savings in consumption were smaller compared to the other methods mentioned.

Figure 13- Amplitude of energy savings obtained using the control method using a frequency inverter with fixed irrigation time (M3) as a function of the spacings analyzed.

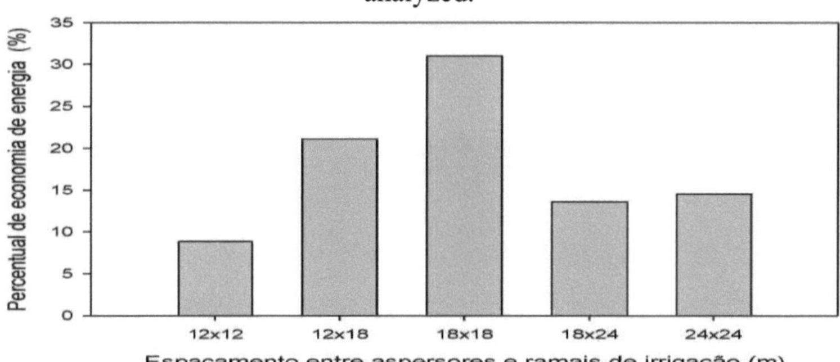

By analyzing all the methods used, the use of the frequency inverter showed the highest savings. A comparative analysis was therefore carried out between the method with a frequency inverter (M3) and methods M1 and M2 to show the percentage of savings. The M3 method showed average savings of 5.17% compared to M1 and 10.08% compared to M2 for the 12 x 12 m spacing, while for the 12 x 18 m spacing the savings were 3.58% compared to M1 and 10.27% compared to M2. For the 18 x

18 m spacing, the savings were between 5.86% and 9.90% compared to M1 and M2 respectively, 0.89% and 7.72% were found for the 18 x 24 m spacing compared to M1 and M2 respectively and for the 24 x 24 m spacing the savings were 2.01% and 8.54% compared to M1 and M2.

Arslan and Sahib (2016) in their study found a reduction in energy consumption with the use of variable speed drives in a pumping system compared to the use of other systems that do not change the speed of the system, with the values found being relatively higher than those found in this study.

This behavior regarding the use of the frequency inverter is evident in Figure 14, which shows the total energy consumption required to carry out one irrigation in each of the situations tested. It is possible to see how much energy can be saved in just one irrigation using this method.

Energy savings behaved proportionally to the increase in area in all situations, however, in the 12 x 18 m, 18 x 18 m and 18 x 24 m spacings this behavior occurred up to areas of 7 ha, decreasing the percentage of savings in areas of 10 ha. The same cannot be said for energy consumption and the increase in spacing. For the 1 and 7 ha areas, the highest savings were found for the 18 x 18 m and 18 x 24 m spacings respectively. For the other areas, the greatest savings were found for the 24 x 24 m spacing in all the methods tested.

The three control methods used showed that the highest energy savings were achieved with the 24 x 24 m spacing, with savings ranging from 189 kWh using method M2 to 611 kWh using method M3.

Figure 14 - Total energy consumption required to carry out one irrigation for each area size, spacing and flow control method

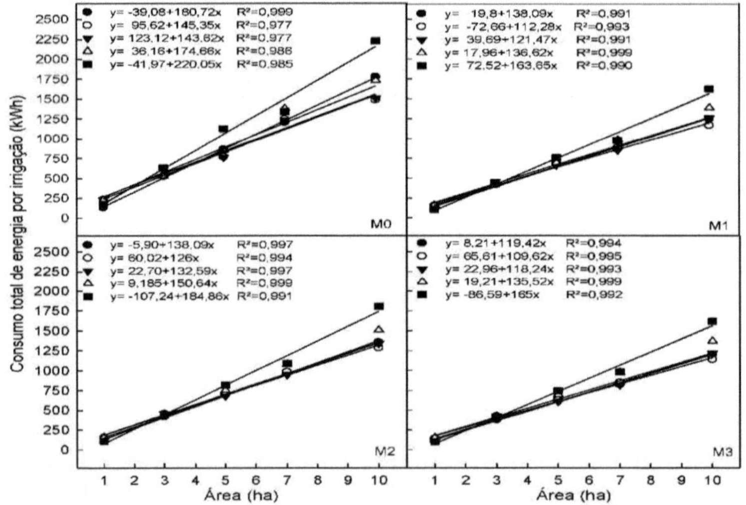

Vieira Júnior et al. (2010), using mathematical simulations in a water supply system in Belém-PA, concluded in their study that there was a 22.7% reduction in

energy consumption with the use of installations with variable rotation in the motor pumps, when compared to those without variable rotation in the same equipment. Rolam et al. (2016), studying SANEPAR's water pressurization system in Telemaco Borba-PR through simulation, found a 30.66% reduction in electricity consumption (kWh), with variable motor pump rotation, compared to its nominal rotation speed under the same conditions. These figures are in line with the results found for this study.

In addition, according to Vieira Júnior et al. (2010), using methods that alter the rotation of the pumping system does not only save energy consumption, it also eliminates the use of valves, high maintenance costs and simplifies assembly of the system's piping. The variable speed flow control method extends the service life of the system's pipes, fittings and motor pump by eliminating water hammer and cavitation.

CONCLUSION

It can be concluded from this proposal that there is a need to verify energy savings by using flow control methods for conventional sprinkler irrigation systems.

Unit energy consumption (kWh mm^{-1} ha^{-1}) in irrigation systems regardless of the spacing between sprinklers shows a tendency to decrease as the irrigated area increases, and total energy consumption per irrigation shows a linear increase as the area increases.

The unit energy consumption increases as the spacing between sprinklers increases. The 12 x 12 m spacing has the lowest energy consumption values and the 24 x 24 m the highest.

The average range of savings was approximately 25% of electricity consumption with the application of the control methods, considering that both the electric motor and the pump are operating within their operational characteristics.

The spacing between sprinklers of 18 x 18 m shows the greatest range of energy savings compared to the other spacings. The spacing with the highest energy savings is the 24 x 24 m spacing, closely followed by the 18 x 18 m spacing.

The M3 flow control method has the lowest consumption and the greatest energy savings compared to the other methods tested. Method M2 has the lowest savings.

The spacing with the lowest energy savings is 18 x 24 m in all the methods tested.

The maximum variation in energy consumption between the flow control methods tested is 10.27% using method M3 compared to M2.

BIBLIOGRAPHICAL REFERENCES

AMERICO, M. **Electronic drive**: advanced guide. Eletrobrás/Procel. Rio de Janeiro, December 2004.

AQUINO, R. R. B.; LINS, Z. D.; ROSAS, P. A. C.; CORDEIRO, L. F. A.; RIBEIRO, J. R. C.; TAVARES, I. A.; AMORIM, P. S. **Energy efficiency in flow control methods**. In: VIII International Conference on Industrial Applications - VIII INDUSCON. Popos de Caldas/ MG. 2008.

ARAUJO, A. C. D.; NETO, J. D.; LIRA, V. M.; LIMA, V. L. A. Evaluation of electricity costs in the context of operation and maintenance of public irrigation projects. **Enciclopédia Biosfera**, Centro Científico Conhecer - Goiania, vol.7, N.12; 2011.

ARAÚJO, J. A. B.; SERAPHIM, O, J.; SIQUEIRA, J. A. C.; and NETO, F. A. **Evaluation of a sprinkler irrigation system using a frequency inverter**. In: Encontro de energia no meio rural, 5. Campinas- SP. 2004.

ARSLAN, S.; SAHIB, A.A. Comparison of energy efficiencies of a small centrifugal pump at constant and variable speed operation. **Tarim Bilimleri Dergisi - Journal of Agricultural Sciences**.vol 22, issue 3, p 444-454. 2016.

AZEVEDO NETTO, J. M. **Manual de Hidráulica**, 8 ed. Edgard Blucher Ltda, Sao Paulo, 1998.

BARUTQU F.; LAMADDALENA N.; FRATINO U. Energy saving for a pumping station serving an on-demand irrigation system: A study case. In **Agricultural production farm and personal training center, Water** saving in Mediterranean agriculture and future. **Research needs**

options Mediterranéennes; serie B. Etudes et Rescherches.n. 56, vol. 1. p. 367-379, 2007.

BERNARDO, S; SOARES, A.A.; MANTOVANI, E. **Manual de irrigapáo**. 8. ed. Viposa: UFV, 2006. 625P.

BEZERRA, S. T. M.; SILVA, S. A. **Automation and control. In Pumping systems**: energy efficiency. Gomes. H. P. (Organizer). Joao Pessoa: Editora Universitária UFPB, p. 206-1250. 2009.

BRIDI, A.B. **Evaluation of energy consumption in pumping system flow control using mechanical valves and frequency inverters - laboratory experiments and case study**. 138 p. Dissertation (Master's degree in Building and Environmental Engineering) - Federal University of Mato Grosso. Cuiabá. 2013.

CAMBOIM, W.L.L.; SILVA, S. A.; GOMES, H. P. Application of Fuzzy techniques to control pressure in water supply systems. **Revista Engenharia Sanitária e Ambiental**. vol.19 n 1. Rio de Janeiro. jan-mar. 2014.

CAMOIRANO, R.; DELLEPIANE, G. Variable frequency drives for MSF desalination plant and associated pumping stations. **Desalination** 182,53-65. Santa Margherita, Italy. May. 2005.

CAMPANA S.; OLIVEIRA FILHO D.; SOARES A.A.; OLIVEIRA R. A.; HERMSDORFF W. Frequency inverters: an alternative for rationalizing the use of electricity in central pivot irrigation systems. In **Encontro de**

Energia no Meio Rural. 3. Campinas (2000).

CARVALHO, J. A.; OLIVEIRA, L. F. C. **Pumping installations for irrigation - Hydraulics and** energy **consumption**. Lavras, Editora UFLA, 2008.

COBO, M.T. C.; DÍAZ, J.A. R.; MONTESINOS, P.; LUQUE, R. L.; POYATO, E. C. Low energy consumption seasonal calendar for sectoring operation in pressurized irrigation networks. **Irrigation Science**, v. 29, n. 2, p. 157-169, 2011.

CÓRCOLES, J. I.; TARJUELO, J. M.; MORENO, M. A. Pumping station regulation in on-demand irrigation networks using strategic control nodes. **Agricultural Water Management**, v. 163, p. 48-56, 2016.

DOORENBOS, J.; KASSAM, A.H. **Efeito da água no rendimento das culturas**. Campina Grande: UFPB, 1994. 306p. (FAO Studies: Irrigation and Drainage, 33).

DUTRA, R.F.; ALTAFINI, C.R. Improvements in a municipal pumping system aiming at energy efficiency - Case study. **Scientia cum industria**, v.2, n.1, p.1-9, 2014.

FARIA, L. C.; COLOMBO, A.; OLIVEIRA, H. F. E.; PRADO.P. Simulation of irrigation uniformity of conventional sprinkler systems operating under different wind conditions. **Revista Engenharia Agrícola**. Jaboticabal, v.29, n.1, p.19-27, jan-mar. 2009.

FEITOSA, E. O.; ALBIERO, D.; PRACIANO, A. C,; MACEDO, D.X.S.; CHIODEROLI,C.A. Simulation of the use of wind energy for irrigation in the low-Acaraú irrigated district - CE. **Revista Brasileira de Energias Renováveis**, v. 3, p. 65-79, 2014.

FERNANDES, R. F.; RUSSI, J. L.; DE SOUZA MUSWIECK, B.; BEHRENS, J.; DA SILVA, G. S. Electric Drive System for a Three-Phase Induction Motor (MIT) Applied to Water Pumping. In: VII Salao Internacional de Ensino, Pesquisa e Extensao. **Anais do Salao** Internacional de Ensino, Pesquisa e Extensao - Universidade Federal do Pampa. v.7, n. 2, 2015.

FORTES, M. Z.; TAVARES, G. M.; DOS SANTOS RIGUEIRA, A.; ANTUNES, V. D. S. F.; PEREIRA, L. F. Energy efficiency by adjusting the operating point of centrifugal pumps. **Proceedings of** the 27th Brazilian Maintenance Congress. ABRAMAN - Associação Brasileira de Manutenpao, Rio de Janeiro. v. 1. p. TT-015. 2012.

FRIGO, J. P.; BOAS, M. A. V.; FRIGO, E. P.; HERMES, E.; TESSARO, E. Day and night irrigation in a conventional sprinkler system in Palotina-Pr. **Irriga**, v. 18, n. 2, p. 318, 2013.

GROSS, J.A.; CASSOL, R. Rainfall anomaly index in the State of Rio Grande do Sul. **Ambiência - Revista do Setor de Ciências Agrárias e Ambientais.** v.11 n.3. set./dez. 2015.

GUIMARÄES, R. **Electrical, mechanical and hydraulic behavior of a pumping system from an energy efficiency perspective**. 173 f. Dissertation (Master's in Electrical Engineering) - Federal University of Uberlândia. Uberlândia. 2008.

HADDAD, J.; VIANA, A.N.C.; SANTOS, A.H.M.; BORTONI, E.C.; YAMACHITA, R.A.; MARTINS, A.R.S. Study of rationalization of electricity consumption in water treatment and supply systems. In: GOMES, H. P. (Org.). **Sanitation Systems:** Energy Efficiency. Joao Pessoa: Editora Universitária-UFPB, 2010. Chap. 22. Available at: <http://www.lenhs.ct.ufpb.br/wp-content/uploads/files/Livro_Eficiencia_Energetica.pdf>. Accessed on: September 28, 2017.

JAMES, K.; GODLOVE, C.E.; CAMPBELL, S.L. **Water and Energy - Taking advantage of untapped water and energy efficiency opportunities in municipal systems**. Alliance for Energy Conservation. 159p. 2002.

KÖPP, L. M.; PEITER, M. X.; ROBAINA, A. D.; TOESCHER, C. F. Characterization of pumping stations in irrigated rice crops from Rio Grande do Sul, Brazil. **Agricultural Engineering**, $36(2)$, 342-351. 2016.

KÖPP, L. M. **Performance indices for pumping stations in irrigated rice fields**. 107 p. Thesis (Doctorate in Agricultural Engineering) - Federal University of Santa Maria. Santa Maria, 2015.

LAMADDALENA, N.; KHILA, S. Energy saving with variable speed pumps in on-demand irrigation systems. **Irrigation science**, v. 30, n. 2, p. 157-166, 2012.

LIMA, A. C.; GUIMARÄES JR.; SEBASTIÄO. C.; FIETZ, C. R.; CAMACHO, J. R. Evaluation and analysis of energy efficiency in center pivot irrigation systems. **Revista Brasileira de Engenharia Agrícola e Ambiental**, v.13, n.4, p.499-505, 2009.

LIMA, A. S.; ZOCOLER, J. L.; JADOSKI, S.O.; MAGGI, M. F.; AMARANTE, R.R. Operational Performance in Central Pivot Pumping using a Frequency Inverter. **Brazilian Journal of Applied Technology for Agricultural Science**, Guarapuava-PR, v.8, n.2, p.51-61, May/Aug. 2015.

MANTOVANI, E.C.; Bernardo, S.; Palaretti, L. F. **Irrigation: principles and methods**. 3.ed. Viçosa: UFV, 355p. 2009.

MAROUELLI, W. A.; SILVA, W. L. C. Selection of irrigation systems for vegetables. **Embrapa Hortalipas-Technical Circular (INFOTECA-E)**, 1998.

MARTINS, C. A. S.; REIS, E. F.; Passos, R.R.; Garcia, G.O. Performance of conventional sprinkler irrigation systems in corn (*Zea mays* L.) **Revista IDESIA** (Chile), Vol 29, N°3, p 65-74, Sept/Dic, 2011.

MESQUITA, R. P. **Technical-economic feasibility analysis for the application of frequency inverters in low-power pumping systems.** 99 p. Dissertation (master's degree in Mechanical Engineering), Guaratinguetá Campus Engineering Faculty, Paulista State University. Guaratinguetá, Sao Paulo. 2009.

MORAES, M. J.; VIEIRA, G. H.; DE OC SCARCELLI, R. Demand-side management in water pumping for irrigated perimeters. **Revista Brasileira de Engenharia Agricola e Ambiental-Agriambi,** v. 15, n. 9, 2011.

MORAES, M.J.; FILHO, D. O.; MANTOVANI, E.C.; MONTEIRO,

P.M.B.; MENDES, A.L.C.; DAMIÄO, J. H. A. C. Automation in a center pivot irrigation system to save electricity. **Revista Engenharia Agrícola**, Jaboticabal, v.34, n.6, p. 1075-1088. nov/dez. 2014.

MORENO, M. A.; DEL CASTILLO, A.; MONTERO, J.; TARJUELO, J. M.; BALLESTEROS, R. Optimization of the design of pressurised irrigation systems for irregular shaped plots. **Biosystems Engineering**, v. 151, p. 361373, 2016.

MORENO, M.A.; CÓRCOLES, J.I.; TARJUELO, J.M.; ORTEGA, J.F. Energy efficiency of pressurised irrigation networks managed on-demand and under a rotation schedule. **biosystems engineering** 107, p 349 - 363. 2010.

MOUSAVI-AVVAL, S.H.; RAFIEE, S.; JAFARI, A.; MOHAMMADI, A. Optimization of energy consumption for soybean production using Data Envelopment Analysis (DEA) approach. **Applied Energy** 88, p 3765-3772. 2011.

Natural Resources Conservation Service- NRCS. **Variable Speed Drive (VSD) for irrigation pumping. Natural Resources Conservation Service**, Water Management Technical Note No. 1, Washington, US, pp. 1-44. 2014.

Natural Resources Conservation Service-NRCS. **Variable Speed Drive (VSD) for irrigation pumping. Natural Resources Conservation Service**, Engineering Technical Note No. MT-14, Montana, US, pp. 1-36. 2010.

NAVARRO NAVAJAS, J.M.; MONTESINOS, P.; POYATO, E. C.; DÍAZ, J.A.R. Impacts of irrigation network sectoring as an energy saving measure on olive grove production. **Journal of Environmental Management**, 111, 1-9. 2012.

OLIVEIRA FILHO, D.; CAMPANA, S.; SOARES, A. A.; OLIVEIRA, R. A. Study of the behavior of electrical and hydraulic parameters in a central pivot irrigation system using frequency inverters to rationalize the use of electrical energy. In: Encontro de energia no meio rural, 3, Campinas-SP. 2000. **Proceedings**... available at:http://www.proceedings.scielo.br/scielo.php?script=sci_arttext&pid=MSC0 000000022000000100017&lng=en&nrm=abn. Accessed on: August 1, 2016.

PANESI, A. R. Q. **Fundamentals of energy efficiency: industrial, commercial and residential**. Sao Paulo: Ensino profissional Editora, 2006. 189p.

PAULINO, J.; FOLEGATTI, M.V.; ZOLIN, C. A.; SÁNCHEZ-ROMÁN, R. M.; JOSÉ, J. V. Situaçao da agricultura irrigada no brasil de acordo com o censo agropecuário 2006.**Revista Irriga**, Botucatu, v. 16, n. 2, p. 163-176, abril/junho, 2011.

REIS, M. M.; SANTOS, L. D. T.; OLIVEIRA, F. G.; SANTOS, M. V. Irrigation of tropical pastures: challenges and perspectives. **Unimontes Científica**, v. 19, n. 1, p. 178-190, 2017.

ROBAINA, A. D. **Agricultural hydraulics.** Santa Maria: Department of

Rural Engineering, 2015, 98 p. Lecture notes.

ROCAMORA, C.; VERA, C.; ABADIA, R. Strategy for efficient energy management to solve energy problems in modernized irrigation: Analysis of the Spanish case. **Irrigation Science** 31: 1139-1158. 2013.

RODRÍQUEZ DÍAZ, J. A.; CAMACHOPOYATO, E.; BLANCÓPEREZ, M. Evaluation of water and energy use in pressurized irrigation networks in Southern Spain. **Journal of Irrigation and Drainage Engineering** 137:644650, 2011.

RODRÍGUEZ DÍAZ, J. A.; LÓPEZ LUQUE, R.; CARRILLO COBO, M. T.; MONTESINOS, P.; POYATO, E. C. Exploring energy saving scenarios for on- demand pressurised irrigation networks. **Biosystems Engineering**, v. 104, n. 4, p. 552-561, 2009.

RODRÍGUEZ GONZÁLEZ, M; SANTANA SOTOLONGO, M; BROWN MANRIQUE, O; ALONSO DE LA PAZ, F. Intermittent flow irrigation associated with onion cultivation and its efficiency in soil and water losses by runoff. **Revista Ciencias Técnicas Agropecuarias**, *22*(4), 50-54. 2013.

ROLAM, M. E.; DE OLIVEIRA NETO, A. G.; BREDA, G. T.; FERREIRA, J. A. M.; WAGNER, L. F. Reduction of energy consumption in water pressurization systems, with speed variation in the motor-pump set. **Revista TechnoEng-ISSN 2178-3586**, v. 1, n. 13, 2017.

SAIDUR, R.; MAHLIA, T. M. I. Energy, economic and environmental benefits of using high-efficiency motors to replace standard motors for the

Malaysian industries. **Energy policy**, v. 38, n. 8, p. 4617-4625, 2010.

SAIDUR, R.; MEKHILEF, S.; ALI, M. B.; SAFARI, A.; MOHAMMED, H. A Applications of variable speed drive (VSD) in electric motors energy savings. **Renewable and Sustainable Energy Reviews**, v. 16, n. 1, p. 543550, 2012.

SÁNCHEZ C. R.; L. SIRGADO, J. L.; PEÑUELAS, F. L.; VILLASANTE, A. L.; SINOBAS, L.R.; LIÓN, G.C. Consumo de energia con bombas de velocidad variable. In: "XXVI Congreso Nacional de Riegos", June 24 - 26, 2008, Huesca, Madrid. **Anais** XXVI Congreso Nacional de Riegos. 2008.

SANTOS, L. D. P. S.; SOARES, A. K.; da SILVA, S. S. Hydroenergetic efficiency in water distribution systems: a theoretical discussion. **Blucher Engineering Proceedings**, v. 3, n. 2, p. 383-392, 2016.

SCHONS, R.L.; ROBAINA, A. D.; PEITER, M. X. Characterization and evaluation of irrigated areas in Rio Grande do Sul. **Revista Tecnologia & Ciencia Agropecuária**, Joao Pessoa, v.8, n.1, p.55-59, mar. 2014.

SECRETARIAT OF AGRICULTURE, LIVESTOCK AND AGRIBUSINESS (SEAPA). 2012. **More Water, More Income**. Available at: <http://www.agricultura.rs.gov.br/conteudo/1032/?Mais_%C3%81gua%2C_M ais_Renda> Accessed on: August 19, 2016.

SOBHY, M. G.; SAMIR, M.; DALIA, M. Dynamic performance application of a variable speed centrifugal pump. In: **National Water**

Research Center, Egypt Fifteenth International Water Technology** **Conference, IWTC 15.** Alexandria, Egypt. 2011.

TARJUELO, J. M.; RODRÍGUEZ-DÍAZ, J. A.; ABADÍA, R.; CAMACHO, E; ROCAMORA, C.; MORENO M. A. Efficient water and energy use in irrigation modernization: Lessons from Spanish case studies. **Agricultural Water Management**, v. 162, p. 67-77, 2015.

TURCO, J. E. P.; RIZZATTI, G. S.; PAVANI, L. C. Cost of electricity in center pivot irrigated bean crop, affected by irrigation management and cropping systems. **Revista Brasileira de Engenharia Agrícola**, v.29, p.311-320, apr./jun. 2009.

TUTTEROW, V.; MCKANE, A. T. **Variable Speed Pumping - A Guide to Successful Applications**. Washington, 2004.

VIEIRA, G. H. S.; MANTOVANI, E.C.; SOARES, A. A.; MONTES, D.R.P.; CUNHA, F. F. Cost of irrigating coffee trees with different types of equipment and area sizes. **Engenharia na Agricultura,** Viçosa, MG, v. 19, p. 53-61, 2011.

VIEIRA JÚNIOR, P.; MESQUITA, A.L.A.; SOUZA, V.H.P DE.; MORAES, G.C. Use of a mathematical model to simulate the operation of a water pumping system. In: GOMES, H. P. (Org.). **Sanitation Systems**: Energy Efficiency. Joao Pessoa: Editora Universitária-UFPB, 2010. Chap. 6. Available at:<http://www.lenhs.ct.ufpb.br/wp-content/uploads/files/Livro_Eficiencia_Energetica.pdf>. Accessed on: December 13, 2017.

WU, P.; LAI, Z.; WU, D.; WANG, L. Optimization research of parallel pump system for improving energy efficiency. **Journal of Water Resources Planning and Management**, v. 141, n. 8, p. 04014094.1-8, 2014.

YACOUBI, S.; ZAIANI, K.; ZAPATA, N.; ZAIRI, A.; SLATNI, A.; SALVADOR, R.; PLAYÁN, E. Day and night time sprinkler irrigated tomato: Irrigation performance and crop yield. **Biosystems Engineering**, v. 107, p. 25-35, 2010.

ZOCOLER, J. L.; LIMA, A. S.; HERNANDEZ, F.B.T.; MAGGI, M.F.; YANO, E.H. Application efficiency of central pivot irrigation using a frequency inverter in the pumping system. In: XXII National Congress on Irrigation and Drainage. Cascavel -PR. 2012. **Proceedings XXII** CONIRD. 2012.

ZOCOLER, J.L.; HERNANDEZ, F.B.T.; SAAD, J.C.C.; CRUZ, R.L. **Payback period for the best-performing hydraulic pump in pumping systems on the green seasonal hourly rate for irrigators**. In: XVIII National Congress on Irrigation and Drainage. July, Sao Mateus, ES. 2008.

yes
I want morebooks!

Buy your books fast and straightforward online - at one of world's fastest growing online book stores! Environmentally sound due to Print-on-Demand technologies.

Buy your books online at
www.morebooks.shop

Kaufen Sie Ihre Bücher schnell und unkompliziert online – auf einer der am schnellsten wachsenden Buchhandelsplattformen weltweit! Dank Print-On-Demand umwelt- und ressourcenschonend produziert.

Bücher schneller online kaufen
www.morebooks.shop

Printed by Books on Demand GmbH, Norderstedt Germany

MIX
Papier aus verantwortungsvollen Quellen
Paper from responsible sources
FSC® C105338
www.fsc.org

Printed by Books on Demand GmbH, Norderstedt / Germany